THE ZEN OF FLYFISHING

PETER KAMINSKY

WORKMAN PUBLISHING • NEW YORK

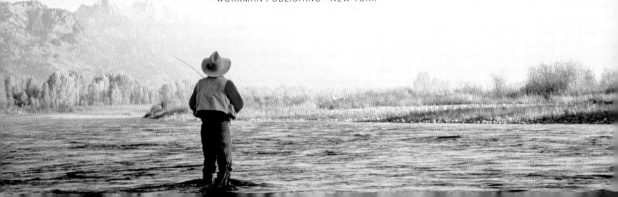

Copyright © 2025 by Peter Kaminsky

Hachette Book Group supports the right to free expression and the value of copyright. The purpose of copyright is to encourage writers and artists to produce the creative works that enrich our culture.

The scanning, uploading, and distribution of this book without permission is a theft of the author's intellectual property. If you would like permission to use material from the book (other than for review purposes), please contact permissions@hbgusa.com. Thank you for your support of the author's rights.

Workman
Workman Publishing
Hachette Book Group, Inc.
1290 Avenue of the Americas
New York, NY 10104
workman.com

Workman is an imprint of Workman Publishing, a division of Hachette Book Group, Inc.
The Workman name and logo are registered trademarks of Hachette Book Group, Inc.

Design by Suet Chong
Cover photo by William Hereford

West Kill in Spring copyright by Steven Weinberg © 2024. All rights reserved. Used with permission.

Additional photography and art credits information is on page 164.

The publisher is not responsible for websites (or their content) that are not owned by the publisher.

Workman books may be purchased in bulk for business, educational, or promotional use. For information, please contact your local bookseller or the Hachette Book Group Special Markets Department at special.markets@hbgusa.com.

Library of Congress Cataloging-in-Publication Data is available.

ISBN 978-1-5235-2453-2

First Edition April 2025

Printed in China on responsibly sourced paper.

10 9 8 7 6 5 4 3 2 1

For Idella

Contents

Preface viii

PART I
THE ART OF FLY CASTING 1
Brief Harmony 4
In Praise of the Quarter Cast 9
A Music Lesson 14

PART II
THE TROUT AND THE FLY 19
The Dry Fly 22
A Choice 33
The Nail in the Coffin 40
Welcome to the Big Leagues 50

PART III
TO EAT (OR NOT) 57
In Pursuit of Happiness 60
Too Fishy? 66
Of Cooks and Casters 73

PART IV
IN PRAISE OF OTHER FISH 79
The Newcomer 86
The Day My Bucket List Kicked the Bucket 92
A Lucky Misfire 99
The All-American 104

PART V
ALWAYS LEARNING 111
The Child Is Father to the Man 115
The Pot No Longer Calls the Kettle Black 122
The Home Game 127
Old 134

PART VI
THE ENDLESS NOW 137
Birds 140
The Light 146
High in the Andes 151
The Persistence of Memory 158

Photography and Art Credits 164 • *Acknowledgments* 165 • *About the Author* 166

*Angling is extremely time consuming.
That's sort of the whole point.*

THOMAS McGUANE
The Longest Silence: A Life in Fishing

An' I have no sense of time.

BOB DYLAN
"Stuck Inside of Mobile
with the Memphis Blues Again"

Preface

In some ways, I see my life as divided into two parts: before flyfishing and after flyfishing. I didn't grow up as a flyfisher. I wasn't even a fisher. If there was meaning in life, I surely hadn't found it before I first cast a flyrod. I'd been a student radical, a New York cabbie, a pastrami steamer, a pipe organ builder, and a go-go dancer on Mexican television. My introduction to writing professionally was at *National Lampoon* in the 1970s. Each day, that job entailed a couple of hours tossing off "can-you-top-this?" one-liners with a collection of hilarious and profoundly cynical writers. Most of our work happened at an extended happy hour where I learned that there may not be a limit on a young man's martini consumption.

And then, in 1976, on a trip to the Yucatán, I found flyfishing. Or, better to say, it felt more like flyfishing found me. On the shores of the Caribbean, just south of the sun-bleached pyramid of Tulum, I chanced upon a row of thatched-roof cabins on a remote spit of land: the Boca Paila Fishing Lodge. It catered to American flyfishers. I was new to angling at the time and had no idea what flyfishing was. During the previous summer, I'd acquired a spinning rod and spent weekends on the Ashokan Reservoir in the Catskills, fruitlessly drowning baitfish in pursuit of smallmouth bass.

When I first saw flyfishing on that visit to Mexico, I was enchanted by the casts of the American sportsmen; the motion of the line and the bend of the rod was lithe, graceful, poetic. Although I have never considered myself a spiritual person, for the first time in my life I felt a spirit come over me. It was, in the words of the hymn "Amazing Grace," the hour I first believed.

From that day to this, my life had found its center. Flyfishing has been the thing that I return to in the rough times that are part of living, but more wonderfully, it has always enriched me in the happiest times. It has been the foundation of friendships, a measure of the forbearance of my wife, a bond with my children, a source of adventure, and an ever more profound connection with our green and blue planet.

Our sense of time and the ways we experience it go a long way toward explaining the allure of flyfishing. There are, I believe, two ways in which we feel time's passage. The first is in everyday life, full of tasks, obligations, sorrows, and also great joys. Daily existence moves to the rhythm of a persistent and inescapable metronome, *time's passing, time's passing, time's passing.*

The second way that we experience time is in the realm of dreams, legends, and myths—when we are lucky enough to be so absorbed in something that the metronome slows until it ceases to tick. For me, flyfishing always does that, but I don't mean to exclude other paths that people may follow on their own route to timelessness. When you find it, though, the minutes, hours, days—even years—feel as if they were a single moment. And when you wake from this dreamland, there is the sense that no time has passed at all.

In this book, I have tried to capture those moments through my most memorable experiences with flyrod and reel, and through the illuminating insights of other flyfishers, photographers, and artists. To be sure, there are highly prized fish in this world that I have never fished for, and angling Shangri-las as yet unknown to me. Still, no matter what quarry you seek and what waters you cast upon, all flyfishers share an unquenchable thirst for "just one more cast," an irresistible impulse to see what lies in the pool beyond the next bend, and a sense that there may be justice in this world every time a cast summons forth a strike. The anglers among you will recognize the emotions, the encounters, and the people who have filled in the chapters herein. All of us flyfishers are bound together by a spirit that inhabits the waters.

PART I

THE ART OF FLY CASTING

Like a tennis player's well-struck forehand or a golfer's soaring tee shot, a good fly cast is a mixture of timing, technique, and a bit of luck. Casting well is one of those rare arts that is both a physical act and a moment of meditation. When you nail it, you know it.

"When I'm fishing well, my concentration is so intensely focused on the surface of the stream that I enter a kind of trance, from which I emerge startled by some sudden sound or change in light. I'll look up, as if just awakened from a dream, and see a great blue heron taking flight at my approach, the tips of his spindly legs lagging three feet behind his crested head, curled claws still skimming the surface of the water."

LE ANNE SCHREIBER
Midstream

Brief Harmony

Although catching fish is the object of flyfishing, for many of us it is the cast that beats at the heart of the game—a graceful and sensually rewarding gesture. When you are casting well, it's as if nothing else is happening in the world except the intimate connection between you and the fly, even though it may be twenty yards away. At the cast's full extension, the moment when your loop unfurls and there is a slight tug of line against reel, things feel good enough to make a skeptic briefly mystical.

Much flyfishing writing would have you believe that this sense of immanence requires the idyllic setting of a Wordsworth verse. Not true: If I am nailing my cast with the prospect of inducing a strike, the scene could be alongside an abandoned warehouse on the Brooklyn waterfront.

You may, at times, become intoxicated with a spate of good casts. *Yeah, I've got this now. I'll just keep at it exactly this way for the rest of my life.* Savor it, try to learn from it, but know that the feeling won't last forever. Like hitting a baseball or a tennis serve, casting is a subtle and evanescent balance of timing, muscle memory, and focus. Great hitters have slumps. Casters do, too. There will be days when your cast leaps forth from your tip-top like a greyhound on the home stretch. Enjoy those moments, but there will just as surely be days—perhaps more of them—when your loop is as limp as linguine. Slumps are part of the game.

When your casting goes through such a dark passage, take a break. Look at the sky, the water, your fly box. When the mist of confusion starts to lift, visualize a well-delivered cast as the last undulation of a wave lapping the shore. Your cast is the final pulse of that wave; applying just the right amount of force at the outset and bringing it smoothly to a halt calls for a short stroke with a minimal exertion of power, precisely timed. Your hand moves, your arm follows, the rod traces its path in a wider arc. You reverse course. The rod bends and snaps back, pulling the line forward until the momentum expends itself and the fly alights.

There is a Zen feeling that comes over you when things go right. Your body, rod, line, and fly are united in a transcendent few seconds.

Metaphysical speculation aside, in the real world of casting doldrums, the one piece of practical advice that I have consistently been given, and found invaluable, is this: Whatever you are doing, do it more slowly. Works like a charm, every time.

"Those rivers I walk and cast a fly upon create magical moments—and not all revolve around catching a fish. Fishing just happens to be the lure that attracts me to them."

STEVE SCHMIDT
"The Art of Letting Go"

In Praise of the Quarter Cast

What is the most flyfisherly way to angle for trout? Many would say it is with a dry fly. But limiting trout fishing to just one style of presentation does a disservice to the time-honored method of quarter casting a three-fly setup. It's the way that my first mentor, Gene Calogero, taught me to fish the Catskills' Esopus Creek, quite early each morning and again in the late afternoon. To cover our bases, we'd fish a stone fly nymph on point, a caddis on the first dropper, and a mayfly nymph or wet fly up high. That setup was known as a cast, not to be confused with the motion that sends line and fly to a target.

There is something hypnotic and dreamlike about the measured cadence of wading into a river, casting back toward the bank, and watching the spray of water trace three slender arcs, one from each fly, as the cast unfurls in the slanting light of morning and evening. The flies descend into the water, and you follow their drift as the current sweeps the line away from the bank. In the argot of steelheaders—an admirably hard-core clan of fishers—this way of presenting a fly is known as "fishing the swing."

Somewhere in the arc of the fly's journey—on a clock face that often falls between 4:00 and 6:00—a fish might take it for an emerging insect and grab it. If nothing happens, you step a pace or two downstream and repeat the exercise, again and again, falling

Down and across—
old-fashioned, but effective.

naturally into a measured and persistent rhythm, as regular and calming as yoga breathing.

At such times, the babble of flowing water, the flutter of newly hatched mayflies, the whooshing of swallows chasing them, the call of birds in the forest, the ruckus of a deer crashing through the brushy understory, and, underscoring everything, the susurrus of the wind blend into a piece of planetary music somewhere between a work song and a lullaby. Have no worries that you will trance out and miss a strike; if a fish is disposed to accept one of your offerings, it will do it with authority. You will feel the jolt, and, roused out of your casting reverie, you'll take up the bracing business of fighting the fish.

A sense of direction is helpful.

> "My father was very sure about certain matters pertaining to the universe. To him all good things—trout as well as eternal salvation—come by grace and grace comes by art and art does not come easy."

NORMAN MACLEAN
A River Runs Through It and Other Stories

A Music Lesson

The year that our daughter Lily turned twenty, she spent a few days on a Pennsylvania trout stream learning to cast with one of America's most accomplished anglers, Cathy Beck, who was a generous and patient teacher. This was by way of preparation for a father-daughter birthday trip to New Brunswick and a few days on the Restigouche and the Miramichi—two fabled Atlantic salmon streams.

But fables don't always match up to reality. It was an unseasonably hot summer, the rivers were low, and the salmon on our first stop, the Restigouche, hadn't put in much of a showing. After a few days, we'd had our fill of not catching fish. We threw in the towel and detoured to the coast, overnighting at a sweet country inn and restaurant in Caraquet, famous for its creamy, plump oysters. We spent the afternoon of our arrival in a nearby forest with a mushroom hunter, gathering an impressive haul of chanterelles and black trumpets that the chef served with seared scallops. For dessert, a cake bejeweled with wild strawberries.

We were primed for our next stop at Pond's Resort on the Miramichi, a hallowed stream that had been at the top of my wish list since 1978, when I first read a *New York Times* Outdoors column by Red Smith about baseball Hall of Famer Ted Williams called "Teddy Ball Game on the Miramichi River." Williams fished that river for Atlantic salmon every year. I feared that even the

Splendid Splinter would have struck out during our visit: more hot weather, low water, and, apart from one wayward leaper, no salmon in sight. We did, however, dine on a supernal planked salmon fillet, served with fiddlehead ferns.

Where, I wondered, had the chef laid his hands on fresh salmon?

Marcia Ponds, who was, at the time, the wife of the proprietor, joined us and offered to take Lily on the water the next day. Marcia specialized in instructing women who were new to the sport. She employed an ingenious method of teaching a proper casting stroke. Her lesson was highly eccentric but, although it sounded loopy, proved to be an effective instructional tool.

She would gather her class for a lesson on the lawn behind the lodge. Pupils were given a long-stemmed wine glass with a standard five-ounce pour of red wine. Next, Marcia would press Play on a boom box and out would come the first notes of "Hold Me, Thrill Me, Kiss Me" by Mel Carter, a syrupy 1960s Top 40 hit often favored as a first-dance song for the bride and groom at schmaltzy weddings.

Each student was told to hold the wine glass by its stem and, keeping time to the music, retract her arm and then move it forward without sloshing the wine. The total travel distance of the glass and its contents was about three feet—the same interval your casting arm would move for a thirty-foot cast. Your motion had to be smooth with a crisp stop. If you could do that without spilling any wine, you had achieved a well-executed casting stroke.

It was rather simple choreography. Each phrase of the lyrics required the flyfishing aspirant to initiate the motion of a back cast followed by a front cast on the next phrase, repeating for a series of false casts:

>*Hold me* (back cast and slight pause)
>*hold me* (front cast and slight pause)
>
>*And never* (back cast and slight pause)
>*let me go* (front cast and slight pause)

. . . and so on, until finally releasing the fly line at the conclusion of the verse. The string section swells. Mel Carter pleads, *Never, never, never let me go*. Go ahead and drink the wine.

On Wildhorse Creek, Idaho.

PART II

THE TROUT AND THE FLY

For most of its history, flyfishing has largely meant trout fishing (with an occasional foray into the aristocratic pursuit of Atlantic salmon). The variety of trout flies, the ingenuity and inspiration behind their creation, and the passion of their partisans has filled thousands of pages. Some flies are admirably restrained and simple in design. Some are bewilderingly complicated. All of them have caught trout. Take your pick.

> "As far down the long stretch as he could see, the trout were rising, making circles all down the surface of the water, as though it were starting to rain."

ERNEST HEMINGWAY
"Big Two-Hearted River"

The Dry Fly

Ask a hundred trout fishers to select the pinnacle of the flyfishing experience, and I believe all one hundred would have the same answer: *a visual eat by a feeding trout*. If you've read any flyfishing literature, then you will no doubt have come across rhapsodic descriptions of trout rising to newly hatched mayflies. I have written more than a few such fulsome passages myself.

In part, the allure of the dry fly goes back to flyfishing's roots as a pastime of the landed gentry of the British Isles. As in all things pertaining to the British upper classes, there arose a code of acceptable comportment that an angler was honor-bound to follow. The celebration of deceiving a trout with the dry fly in a sinuous, ever-flowing chalk stream was understood by all to be flyfishing's supreme moment . . . provided that you fished in the manner prescribed by Frederic M. Halford in his 1889 work *Dry-Fly Fishing in Theory and Practice*. That meant your cast would be upstream and only to a rising fish:

> *"Some dry-fly fishermen are such purists that they will not under any circumstances cast except over rising fish, and prefer to remain idle the entire day rather than attempt to persuade the wary inhabitants of the stream to rise at an artificial fly, unless they have previously seen a natural one taken in the same position."*

This technique was well-established flyfishing orthodoxy when I took up the sport in the 1970s. Chuck-and-chance casting or fishing subsurface were judged by dry-fly devotees to be crude and inelegant.

Fishing according to the canonical method on freestone Catskill streams, I caught my share of trout on dry flies. But I soon learned that flyfishing had evolved since Halford's day and that the dry fly presented *downstream* is pretty much the only way to catch super-selective risers on such rivers as the Delaware, or, farther afield, the Missouri or Patagonia's Arroyo Pescado. And if nothing is doing up top, nymphing will always outscore the dry fly. Still, Halford was quite right that there is no experience in trout fishing as viscerally exciting as casting to a trout as it rises in rhythm, taking duns. Seeing your quarry, observing it in the wild, planning your approach—all culminating in the moment when things come together and you have a trout on the line—is trout fishing's biggest big payoff.

In the century and a half since Halford evangelized for the dry fly, the pursuit of fish with a flyrod has expanded all over the world and includes many more species than the trout and salmon of those earlier years in the United Kingdom. But the thrill of stalking a fish and readying your cast is the same whatever the species: tailing bonefish with the morning sun glinting off their waggling tails; the willowy black brushstrokes of a permit's fins as it leisurely devours tiny crabs on a sandy flat; the sudden turn of a tarpon cutting away from the pack to chase a fly; the bulging bow wave and the explosion of a largemouth, its mouth open wide, as it crushes a popping bug.

Visual encounters like these ignite the food-hunting genes we inherited from our Stone Age ancestors, the same heritage that moved Halford when he felt goose bumps at the sight of a handsome brown trout in the shade of a flowering chestnut tree on the banks of a Hampshire stream. With the regularity of a stopwatch, the trout would drift downstream for a few feet to swallow a meaty drake, then return to its station, ready to pick off another target. It's a performance that never gets old.

The ring of the rise is Nature's way of telling you, "Put your fly on the bullseye."

Left to right: From dun, to vise, to classic old-school dry fly.

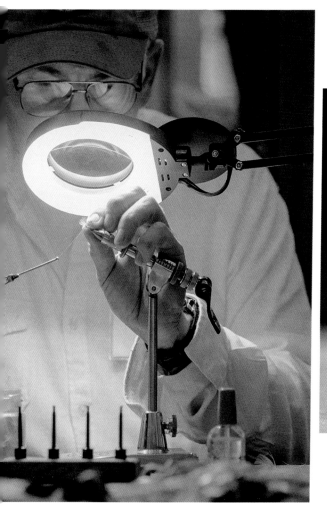

"I look into the plastic Dixie Cup that is my fly box and think about all the elements I should consider in choosing the perfect fly: water temperature, what stage of development the bugs are in, what the fish are eating right now. Then I remember what a guide told me: 'Ninety percent of what a trout eats is brown and fuzzy and about five-eighths of an inch long.'"

ALLISON MOIR

"Love the Man, Love the Fly Rod"

A full-on hand and then a spinner fall blankets the stream.

There's always a next fly.

A Choice

What is the best fly? Just as there are stages in the life cycle of a mayfly, there are epochs in every angler's fly selection. We go through phases of wanting to match the hatch with great precision. On the other hand, we also reach a point where, apart from the assumption that a fly bears at least a general resemblance to the natural, the question is more how an artificial behaves in the water. Does its action resemble that of a real insect? Most important, does it catch fish?

If you asked me to choose one, I'd probably come down firmly on the side of both. This may not make logical sense, but it does make fishing sense. For selective trout on spring creeks and glassy pools, I believe size and silhouette are critical, so my dry fly pick is the Comparadun—the no-hackle invention of Al Caucci and Bob Nastasi. It rides low in the water and, when seen from below, has the outline of a newly hatched dun. Color doesn't mean nearly as much as the School of Exact Imitation would have you believe. Remember, the angler watches the fly from above, while the trout's view is from below, looking up. It sees a dark outline against a bright sky. I believe that it recognizes a silhouette more than it registers color.

On riffled or pocket water, I like an old-school stiffly hackled fly—say, an Ausable Wulff or an Adams. Hackle, according to John

Atherton's 1951 classic *The Fly and the Fish*, breaks up the light just as a newly emerged mayfly does as it attempts to take to the air for the first time. For small stuff, I always have a Griffith's Gnat handy. To a trout, I suppose it looks like the furious flutter of tiny wings. As for caddis flies, I have caught very few trout when there is a caddis hatch on—I must be doing something wrong—but I have caught many trout by prospecting the water with an Elk Hair Caddis. It is a fairly primitive tie and I have no theory why it works. I only know that it does.

Subsurface refers to flies that are taken as nymphs or emergers (for caddis flies, that would be pupae). This is where I am more concerned with movement than imitation. Trout go for subsurface insects because they are triggered by the action of a nymph or pupa shucking off its case as it prepares to take flight. Like a person trying to wriggle out of a pair of undersized jeans, there's not much else that insects in mid-metamorphosis can do other than finish their clumsy disrobing. That makes them easy targets.

Flies work best with filaments of soft hackle or with dubbing that creates air bubbles, which give the appearance of bugginess to feeding trout. They are not so much replicas of an insect as they are imitations of the choreography of its hatching movements. According to this principle, flies such as the Gold Ribbed Hare's Ear or Pheasant Tail Nymph can be taken for many aquatic insects. These flies have been around for a long time, and for good reason: They produce.

And then there is the Prince Nymph. I caught my biggest sea-run brown trout (twenty-six pounds, twelve ounces) and my biggest rainbow (a potbellied twenty-pounder) on a size-12 Prince Nymph. I am not sure what the fish thought they were consuming. The rainbow must have taken it for a freshwater shrimp, because it lives in a lake full of them. And the sea-run brown? Nobody knows why spawning run salmonids strike; pure orneriness is often an explanation, and I think it is as good as any.

To sum it all up, a trout is liable to take anything that looks like food, especially if the fly resembles the appearance and movements of what it's chasing at a particular moment. If you have success with a particular fly, keep using it. Having faith, especially if it is born of experience, is preferable to fishing according to a scientific discourse. Rather than keeping an enormous selection of flies on hand, you might opt for a small assortment that covers many bases. If I had to pick a handful, then, I'd go with the Comparadun, Ausable Wulff, Elk Hair Caddis, Griffith's Gnat, Prince Nymph, Gold Ribbed Hare's Ear, and the Wooly Bugger. You can trout fish anywhere in the world with these, and rest assured you will catch fish if they are there to be caught.

"An everlasting round of money getting contributes nothing to the public morals. On the contrary it crowds our insane asylums . . . . Show me a people who love innocent out-of-door recreations, says the philosopher, and I will show you a community where peace and virtue dwell."

ROBERT BARNWELL ROOSEVELT,
Thirteenth Report of the New York Fish Commission

"In that beautiful place, the impossibility of knowing filled me with an irrational flush of happiness ('a springing up of happiness equal to a hymn, to loosing a cloud of pigeons, to the river singing')."

JULIO CORTÁZAR
"The Distances"

The Nail in the Coffin

Art Flick's *Streamside Guide* was the first book that I regarded as a sacred flyfishing text. In clear, direct prose, Flick simplified the progression of mayfly hatches that unfolds on the Catskills' Schoharie Creek. He narrowed the choice of flies down to just a few of the likeliest candidates instead of the scores of species you could spend a lifetime trying to imitate. For the overwhelmed neophyte, it showed me that fly selection was a hill I could climb.

Still, I have a bone to pick with the "Laird of the Westkill," as Flick was known, for setting me off on my futile decades-long quest for a trout bonanza that he so vividly conjured, and I have chased, ever since: big brown trout feeding on coffin flies, the ivory-bodied, black-winged spinners of the green drake.

Coming in the later innings of the cycle of hatches, the green drake—which is a very large mayfly—will intermittently pop up on a quiet stretch of water, bringing to mind a drunk startled out of bed the morning after. The newly emerged mayfly will flutter its wings with great commotion and essay a few clumsy attempts to lift off, which, in all fairness, is a new skill for a creature that has spent its life to that point as a wingless insect in a silty river bottom.

Because the fly is so big, it can attract trophy fish not otherwise motivated to come up for a smaller mayfly, preferring instead to consume nymphs or small baitfish. Nevertheless, because the hatch is so sporadic, it rarely triggers the rhythmic rise of many trout.

The big green drake event is not, however, this desultory emergence of the duns, Flick said; instead, it is the spinner fall, when a few days' worth of drakes have assembled in their mating regalia of black and white and gather in a growing congregation over the stream. If you are on the water when this happens, then hour by tantalizing hour, you will watch in hope that the cloud of copulating coffin flies finish their descent and trigger rising fish.

I have seen this movie a hundred times but never made it to the final scene. One day, however, stands out in memory because, after whiffing on her driving test in Brooklyn, my teenage daughter Lucy reasoned that she might stand a better chance if she tried her luck with a more easily impressed tester in the Upstate hinterlands. No luck. But we were near the West Branch of the Delaware. So, as consolation, we fished. After a good half hour of working a bank feeder, Lucy delivered an elegant left reach cast with a size-16 Elk Hair Caddis on 5X tippet. It seduced a sixteen-inch wild brown, which, when it came to the net, was made extra memorable by its impressive kype.

The afternoon went on and we watched lustfully as an hours-long green drake mating swarm descended lower and lower. Barely a nanosecond before dark, when the dense formation of coffin flies were about to kiss the water, the gentlest of breezes came up and the big spinners dispersed.

"Goddamn them!" I grouched, vowing to give up this empty and never-ending quest for trout fishing's El Dorado—at least until the next time I come across a green drake mating dance.

A spinner in silhouette (left) and on the stream (right).

"I find in rivers a half-seen world of fleeting forms that tempt and sometimes ensnare the heedless soul. Rivers and streams are sinuous, insinuating creatures, untamed. Not one of them has sure limits."

HENRI BOSCO
Malicroix

Riffle, pool, riffle—the age-old river song.

"Fishing teaches a stern morality; inculcates a remorseless honesty. The fault may be with ourselves. 'Why do I go on missing at the strike? . . . If I had more delicacy in casting, more accuracy, if I had fished finer, should I not have done better? And the answer is—Yes!'"

VIRGINIA WOOLF
The Moment and Other Essays

Welcome to the Big Leagues

You won't forget your first big fish. Most likely you won't land it, either. A big fish is a new and fiercesome level of the game, sure to catch you off guard. My big-fish baptism happened on Armstrong Spring Creek, one of the pristine feeders of the Yellowstone just outside Livingston, Montana.

In those days, before Robert Redford's film version of *A River Runs Through It* sparked a stampede of anglers to Montana's bountiful trout-filled waters, all you needed to do to fish Armstrong was show up at the modest ranch house residence of the O'Hair family. Using the top of Mrs. O'Hair's washing machine as your writing surface, you'd fill out a check for thirty-five dollars, hand it over, and head to the stream. It was like walking into the cover of your favorite issue of *Fly Fisherman* magazine. Life rarely gives you the opportunity to step into your fantasy.

I wadered up and eased into the clear waters. Grassy weeds swayed in the current. Trout were everywhere. Pale Morning Duns, a dependable Montana hatch, fluttered on the surface, drying their wings before taking flight. The fish should have been feeding, but the purple and yellow flies floated by unmolested. It was early in the hatch, I reasoned. The fish were probably gobbling emergers. No doubt, as I made my way in the current, disturbing the river bottom with my boots, every trout in sight wised up to me and passed the warning to their cousins upstream.

It ain't over 'til it's over.

I reached a place where a slough emptied into the stream. Quiet water, the kind that makes you hold your breath. A trout rose, its presence marked by a dimple. I tied on a small, buggy attractor and released the cast. The fly landed above the ring of the rise. A trout took. I came tight. The biggest trout of my still-young fishing life, a two- or three-pound brown, went airborne, shook its head, and, just like that, broke off. I was a bit discomfited but counted that brief connection to the fish as a small triumph.

"Not really a success," my friend Larry Aiuppy reproved from his place on the riverbank. Larry is a formidable fly-fisher who, in addition to having the most delicate touch, can make his way across the swiftly flowing Yellowstone as easily as a bison.

"The complete act," he said, "is to present the right fly, to induce a strike, to bring the fish beneath the bend of the rod and not allow it to take you under banks and dead falls where it can break free, to bring it to hand, to land it, to revive it, and to watch it swim away. Anything less doesn't count."

In other words, there is no such thing as good enough. Carrying a no-hitter into the ninth inning and then giving up the victory on a solo home run is not a semi-no-hitter.

A fish is well caught, or it isn't.

An oxbow bend in the river surely holds big trout.

THE TROUT AND THE FLY

"There will be days when the fishing is better than one's most optimistic forecast, others when it is far worse. Either is a gain over just staying home."

RODERICK L. HAIG-BROWN

Fisherman's Spring

PART III

TO EAT (OR NOT)

All anglers love catching fish.
They feel they've done a
good deed when releasing one.
Still, it's a great pleasure
from time to time to catch dinner.

"With my rod and only one fly, I will catch a small bluefish, eat sushi, chew on some seaweed. Suck on the last of the lemon drops. Morning will come, a sunrise of indisputable hope and renewal. The striped bass will roil in, just for me, and I shall cast, catch, and release these great creatures from the ocean. Later in the day, the Coast Guard will pick me up on my deserted island."

MARGOT PAGE
"The Island"

In Pursuit of Happiness

It should be said that catch-and-release fishing is a fine conservation principle, but it is not a bedrock moral commandment that ennobles the catcher. I have kept no more than a dozen trout in the last forty years; most of those were caught in an isolated lake in the Andes where hardly anyone else fishes. It abounded with pound-and-a-half brook trout that would take on nearly every cast. Catching them was as easy as picking ripe apples off a tree. Their flesh was the vibrant coral color of salmon, and when cooked in a skillet over a wood fire, they were as delicious as food ever can be. Still, the plain fact of the matter is, if you love catching wild fish, there is no way to sustain a fishery while keeping every fish you bring to hand.

Trout are one of the apex predators in rivers and streams. By natural law, there are far fewer of them than the insects and smaller fish that they prey upon. By the same principle there are far fewer lions than there are impala, fewer wolves than elk, fewer alligators than bass. Top predators are an especially vulnerable natural resource and easily degraded. So, if you enjoy the rise, the cast, the take, the bend of the rod, and the song of the reel, catching and releasing is a wise practice. When you hear people getting sanctimonious about it, however, it's worth reminding yourself that at its most Darwinian level, the joy of fishing stems from the successful pursuit of food—that is to say, the killing of another creature.

Fishing, no less than hunting, is born of the same primal and lethal instinct. We may take atavistic pleasure in the struggle that ensues with a fish on the line, but the fish are unwilling partners in this exercise, and there is no question that they suffer. You can release a caught fish—with care to help ensure its survival—but you can't do the same with a deer brought down with a rifle, or a grouse taken on the wing.

As Lee Wulff sagely put it, in explaining catch-and-release, "Game fish are too valuable to be caught only once." Catching makes you happy. Releasing makes it possible for others to experience the same joy. We can be content to leave it at that.

To attract a fish to your fly, you must first understand what it is eating.

Too Fishy?

People who will readily order seafood in a restaurant are often hesitant to cook fish in their home kitchen. I saw this quite clearly at a dinner I attended in Charleston, South Carolina, while working on a story about sustainable fisheries. The bill of fare that evening was all locally caught on hook and line: olive oil–poached amberjack, lemon and rosemary–braised porgy, and roasted black sea bass. The diners, about sixty people, were enthusiastic about the meal, but when I went from table to table to ask if they cooked fish at home, a clear plurality said no. When I asked why, the most common answer was "too *fishy*."

Closer to home, I've encountered similar prejudice directed at one of my favorite flyrod quarries, the pugnacious bluefish. On the morning of a recent Labor Day cookout in my backyard, a friend invited me to fish the rip that sets up on the outgoing tide off Coney Island. I caught a bunch of blues, filleted them, and packed them on ice. Rubbed with oil and seasoned, they cooked up nicely over a wood fire. Despite a prevailing low opinion of bluefish, the guests at our barbecue all agreed that when they're prepared fresh from the ocean, their flavor is pleasingly delicate, especially if you remove the dark meat that runs across their lateral line.

As I found with those blues, there is always a special delight in eating a fish you have caught; it completes the narrative. For me, the determining factor in whether it's okay to kill a fish is the sustainability of the fishery (bluefish are quite plentiful). Striped bass, on the other hand, are more pressured. Keeping just one striper a year—which I do when the fishery can sustain even that minimal harvest—transforms the meal into a sacrament, a form of Communion.

A whole fish, say eight to twelve pounds, roasted in a thick mantle of salt, is as festive as a glazed ham or a turkey for a holiday meal. There is even a bit of drama as you crack open the hardened salty crust, releasing a burst of steam that clears to reveal the moist white flesh of the bass. King salmon, steelhead, lake trout, and redfish will elicit the same lusty *oohs* and *aahs* from your dinner party.

In my early, clueless years of flyfishing I would fill my creel with a half dozen Esopus Creek rainbows caught during the *dorothea* hatch. I smoked the trout at our cabin on the Little Beaverkill, one of its tributaries. The fillets were the main ingredient in a trout salad dressed with homemade mayo. Breakfast was the real luxury: an omelet filled with smoked trout cheeks and garnished with minced chives. It was thoroughly delicious, but after I learned about the precarious state of trout in the ecosystem, I gave up this annual ritual. I never took another trout from that river. If keeping my catch would not have degraded the fishery, I calculate I would have consumed something in the neighborhood of six hundred trout cheeks by now. For a flyfisher, you can't have your trout and eat it, too.

I always always
 always bring a landing net.
  Not today of course.

HAIKU BY CHRIS ENGLE

"The way you drift or retrieve your fly is far more important than either its size or pattern. So before you become submerged in the academic judgment of fly selection, learn the fundamentals of making the fly's actions appeal to fish. How you handle your feathers is the true measure of your angling skills."

RICHARD L. HENRY
"The Drift and Float for Trout"

Of Cooks and Casters

Some years ago, I wrote a story for *New York* magazine covering the lead-up to the opening of Gramercy Tavern, where Tom Colicchio was the inaugural chef. One evening Tom invited a few members of the Gramercy team and me to his Manhattan apartment for dinner. He busied himself at the stove with the finishing touches on his gravy—a term that is understood by all native-born citizens of New York and New Jersey as traditional Italian red sauce. As I wandered around the living room, my eyes lit on some beautiful Red Quills (the male of the Dark Hendrickson) that Tom had tied; they were quite sparse with a clean profile and just-enough hackle to break up a shaft of sunlight.

It struck me then that tying a fly and preparing food are skills that require a highly attuned sense of touch and feel. Recipes for flies, like recipes for dinner, are not the precise instructions that you might find in a shop manual for a piece of machinery. Organic materials vary, and you need to adjust to subtle differences in ingredients, all of which are likewise products of nature—things that were once alive. Fish, after all, take flies that behave like living things. Flies, like food, are often made of organic material (or material that behaves like nature's originals).

Perfect follow-through; stop the rod high.

A few months after that dinner of spaghetti and meatballs, another similarity in the mindset of good cooks and anglers struck me. A creative chef always wonders, "What happens when heat comes in contact with ingredients? Does it char or gently warm? Does it crisp or moisten?" In other words, the written recipe is primarily preparation for the encounter between ingredients and heat, and that requires a bit of adjustment every time you cook. Flyfishing is not that different. Many anglers feel that once they have matched the hatch and executed a good cast, they've fulfilled their part of the fish-catching bargain, but the most successful angler *begins* by visualizing what happens *after* you have sent a fly on its way and placed it where a fish might see it. Does it break up light? Does it undulate like a living baitfish? Does it wriggle like a mayfly discarding its nymph case? How can you deliver a cast and manipulate the line to give a fly the longest possible time to provoke a strike? Successful fishing, much like a well-prepared recipe, really begins with an act of imagination.

Wild Brook trout caught by the author from a lake in remotest Patagonia and cooked with rosti potatoes on a plancha.

"Fishing doesn't actually happen. It just goes on in your head."

ROBERT RUARK
Old Man's Boy Grows Up

PART IV

IN PRAISE OF OTHER FISH

The biggest change in the modern era of flyfishing has been a wider appreciation of fish beyond trout and salmon that will take a fly and fight well. The world of flyrodding has grown: There are new places to go, new people to meet, and, of course, new challenges each fish presents. It's especially nice to discover that some of these quarries have been swimming just beyond our doorsteps.

Snook are built like broad-shouldered linebackers and fight like angry largemouth bass.

"...the silver king swam swiftly and strongly with occasional joyous leaps, helped on by wind and tide, and always seaward..."

A.W. DIMOCK

"The jungle people believe in a sun god who had two children of profound likeness—one the jaguar that rules the forest, and the other the golden dorado that rules the rivers."

KIRK DEETER

"Son of the Sun God"

The Newcomer

Over the last two centuries, flyfishers have spread across the world and added species after species to our sport, but few have achieved a place in the pantheon alongside brown trout and Atlantic salmon. Post–World War II, the pursuit of bonefish, permit, and tarpon on white coral flats added these tropical fish to the elect. Then, in the early years of this century, reports started to arrive from South American flyfishers who had begun to pursue golden dorado in Bolivia and northern Argentina. Soon the word spread, and a new flyroddable fish took its place on the summit alongside the salmonids and the flats fish of the tropics.

Among its virtues, dorado will attack a fly as aggressively as pike. No fish I have ever encountered will as readily and repeatedly go airborne, leaping and thrashing, forcing you to bend your rod this way and that to keep the contest manageably close. When the dorado finally comes to hand, it glints like an Inca sun disk.

I have caught dorado in the grassy wilderness of Argentina's Iberá wetlands, where frequent sightings of capybara, the world's biggest rodents (weighing up to sixty pounds), make for an exotic wildlife bonus to a fishing outing. Closer to Buenos Aires, you can drive about a half hour out of the city and arrive at the delta of the Río de la Plata. It is dotted with islands among the braids of the river, and when the sun spreads its shafts over the scene, the vista is as luminous as a William Turner seascape. Though dorado of the delta rarely run to trophy size, they are close to town and make for a nice day's flyfishing away from the bustling pace of the capital.

My most profound dorado experience was on the Sécure River in the Bolivian headwaters of the Amazon. It is the remotest place I have ever been. To reach it, you fly into the town of Santa Cruz and then take a small plane that passes over coca plantations in the jungle. The undersides of the wings are marked so that the coca growers don't mistake your plane for drug police and shoot at it. At least that's what we were told, and it added to the Indiana Jones vibe of the trip. Then, landing on an airstrip hacked out of the tropical forest, you transfer to the long, slim canoes typical of the region and make your way upstream to the stretch of river inhabited by the indigenous Tsimané, who still hunt with bows and arrows.

On our first angling session, my brother Don and I caught a half dozen dorado that were bigger than any I had yet seen. It was glorious fishing as wolf packs of dorado churned the water, ripping through schools of menhaden-sized *sabalitos*. On the following day, after a productive morning, Don and I waded downstream and got into a few more nice fish. As we began to make our way back to the lodge, the midstream rocks we had noted on our way downriver had disappeared. The water was rising fast. Within an hour, the river came up two meters. The churning, swollen Sécure turned dark and remained unfishable for days. We later learned that there had been an intense cloudburst upstream.

Anglers work their way upstream on the Sécure, home water of the Tsimané people.

On our last morning, we were finally able to try for dorado again, just below the Tsimané village. Fishing was slow, though one member of our small group took a magnificent twenty-pounder. He whooped appropriately. And then, as if on cue, time froze and we were riveted to our spots as a full-grown jaguar stepped into a sunlit space on the opposite bank. The beautiful beast, in its regal mantle of gold and black, momentarily looked our way, then, dismissing us, turned its head and disappeared into the green shadows of the rainforest. We felt as if the jungle was sending a message; it had shown us everything, and it was time to go.

Closer.

Rolling a lead-eyed hackle

back in the riot

of mouth and tail. *Come on, take it!*
  I beg the fish,

then you, *Come on. A little closer.*
  Another cast.

Fish moving fast. Rocks under
  the hull.

That's enough, you hit reverse.

Please, I say. *Just a little closer.*

HENRY HUGHES
A Little Closer

The Day My Bucket List Kicked the Bucket

When I first took up flyfishing, it seemed to me that life would be complete, or nearly so, if I could fish those places that enthralled anglers in the pages of *Fly Fisherman* magazine. I wanted to double haul into the teeth of a Patagonian wind like Joe Brooks; lay down a dry fly on an English chalk stream, its grassy banks as carefully clipped as the fifteenth hole at Augusta; or offer up a few Deceivers just like Lefty Kreh, shoehorning back casts between passing cars speeding alongside an Everglades canal filled with baby tarpon. As a fishing journalist, I was able to fulfill many such dreams: Christmas Island bonefish on Christmas Day, Atlantic salmon in the aquamarine waters of Iceland's Miðfjarðará, tarpon at sunrise in the Marquesas.

It wasn't until the pandemic and the quarantine of 2020 that I realized such lists often aren't much more than esteem-boosting scorecards for anglers, the theory being that the more places you can check off, the closer you come to flyfishing satori. But when the world shut down, a trip to New Zealand for the mouse hatch would have to wait. Big browns slamming salmon flies on the Madison could enjoy the hatch without me getting in the mix. My longtime fantasy of eating an onion sandwich, à la Papa Hemingway, while fishing for black trout in the Pyrenees would be put off for yet another season.

For two years, dream trips were, of necessity, placed on hold. There would be no flying in my flyfishing. But, thankfully, there were 566 miles of New York City's coastline and its rich estuarine waters extending from the end of my street to a hundred miles offshore where the Hudson Canyon descends into bottomless blue depths.

Some of my friends kept boats out on Mill Basin at the end of Flatbush Avenue in the Canarsie section of Brooklyn. From their moorings at Sea Travelers Marina, it was a short trip to the waters of Jamaica Bay, the rip just offshore of the Parachute Jump at Coney Island, the south shore of Staten Island, and over to Sandy Hook, New Jersey. For some reason that I have never fully understood, the nearest thing to a guarantee of fish is in the bay right next to a runway at John F. Kennedy Airport. Quite often the stripers will drive the bait up against the shore. It was rather hard—actually, impossible—to resist the temptation to move closer to them than the six hundred feet enjoined by airport security.

As if they'd read the memo, the fish appeared to know that they would not be molested as they corralled bunker inside the no-boat zone. Inevitably, you'd creep over the line a bit and, dollars to doughnuts, a police patrol boat would motor up from its holding position by the trestle where the A train crosses Broad Channel to Far Rockaway. In the tone of weary disappointment that parents reserve for a child who has let them down, the harbor police would share some variation of "C'mon, guys. You know you can't be here and we know you can't be here, so why are you busting our chops?" Very NYPD.

My special fish entered the scene on November 11, Veterans Day. We had covered all of Jamaica Bay. There was plenty of bait and, under the bait balls, a ton of bass on the fish finder. But they weren't eating. We waited out the tide and still there were no customers. Reluctantly, we began the journey home, hoping to see birds working, but apart from a few aimless terns, nothing was happening. We passed the enormous landfill, now reclaimed as a pretty state park named after Shirley Chisholm, America's first Black congresswoman.

I had given up on the day when my fishing partner, Kurt Schwarz, cut the engine on his Maritime skiff and picked up his binoculars.

"Let me check out the creek that feeds in here. I think I saw birds," Kurt said. He is both an optimist and indefatigable.

"Sure," I said by way of implying, *Let's get this over with*.

"Maybe something there. Hard to tell. Let's motor up the creek, nice and slow."

We eased up the creek at an idle, hardly pushing water or throwing a wake. It felt fishy. If you blocked out the occasional out-of-commission shopping cart tossed up by a storm surge, the scene had the look of an Everglades canal, high banked and maybe sixty feet wide. Enough room for a back cast, but not much of a haul.

"Worth a few casts," Kurt said.

He handed me his rod, an 8-weight with a big foam popper on it. My cast landed on the bank. A little wrist action coaxed it to the water at a break in the weeds. On the retrieve, the popper made slurpy, sloppy burbles.

Whoosh!

The biggest striper I'd seen that year reared back and made for the near bank. Then the far bank. Then under the boat. Then around the stern, forcing me to raise the rod and my arms to clear the outboard. Kurt coached my every move as if it was the first time I'd hooked a good fish. Or maybe he was just cheering me on. Or both.

When I tell this story, I have taken to calling the beautiful striper a fifteen-pounder, but then, angling memory is a magnifying lens. Surely, though, it was a serious flyrod fish and as satisfying as any I have caught after traveling ten thousand miles.

I drove home along Brooklyn's still-shady streets, the leaves a mix of summer green and autumn gold. I stopped on the way for a cold beer and an order of fried calamari with spicy marinara sauce at Randazzo's Clam Bar in Sheepshead Bay, right across the street from the party boat fleet.

As often happens with fishing, a melody began to play on my inner jukebox, a Billie Holiday tune whose lyrics could have been written for this day, laid-back and jazzy:

> *You'll find your happiness lies right under your eyes*
> *Back in your own backyard.*

"It is not possible to describe how this bay swarms with fish both large and small, whales, tunnies and porpoises, whole schools of innumerable other fish, which the eagles and other birds of prey swiftly seize in their talons when the fish come up to the surface, and hauling them out of the water, fly with them to the nearest woods or beach."

JASPER DANCKAERTS

describing a visit to Manhattan, 1679

Journal of Jasper Danckaerts, 1679–1680

A Lucky Misfire

Practically speaking, trout fishing in the Catskills often ends in the first few weeks of October. You might get lucky and tie into a spawning brown at that time of year, although I never have. I'd catch a few fish on Indian summer days, sometimes on the White Gloved Howdy (which imitates the female spinner of the Isonychia Bicolor). It is one of our most prolific and productive hatches. And what a name! It puts me in mind of the glad-handing politician that Chuck Berry evokes in "Nadine": "campaign shouting like a southern diplomat."

But I digress. Along about mid-October, I would take my twenty-gauge Browning and go tramping through the overgrown apple orchards on the hillside in Arkville, New York, pursuing ruffed grouse and woodcock. One crisp autumn morning, I put up a woodcock. I raised my shotgun, but it refused to fire.

"Damn, must have left the safety on!"

I checked the safety setting and marked where the woodcock went down. You'll often get a second flush out of them.

The bird went up again. I swung on it, pulled the trigger, and . . .

"Damn, must have left the safety on again!"

Weird. But I continued breaking through berry tangles and scrambling over blue stone walls, raising a commotion that I hoped would frighten any birds that were holding tight and give me more shots. I had already shot a grouse a few hours before, so I counted

Being There by Bob White

the day as a win. My brother Don was with me, and he had his camera along. He suggested that I break open my shotgun and place my grouse beneath it. It would make for a nice photo, he said.

I retrieved the grouse from my game pocket and lay it directly on a bed of ferns. I opened the Browning to finish the tableau and—dammit—there in the shotgun's improved cylinder I saw the reason for the misfires. What had happened was that after shooting the grouse, I was under the impression that I had put a new shell in the empty chamber. On closer inspection it turned out that, in my rush, instead of a shell I had jammed in a package of Rolaids.

I tell you this by way of saying that much as I enjoyed everything about bird hunting, it just wasn't in the cards that I'd ever become good at it. The Rolaids sealed the deal. Right about then, I discovered saltwater flyfishing out at Montauk Point. It took a few years to get the hang of the bigger rods and ever-fickle winds, but the upshot was that I could add a few more months of flyfishing on either side of trout season without feeling compelled to go to Florida or the Bahamas for bonefish, tarpon, and permit. That wayward pack of Rolaids was, in its way, a blessing.

Breezy Point by Joel Stoehr

"I consider him, inch for inch and pound for pound, the gamest fish that swims. The royal Salmon and the lordly Trout must yield the palm to a Black Bass of equal weight."

JAMES ALEXANDER HENSHALL
The Book of the Black Bass

The All-American

Flyfishing for bass has introduced me to an America I might have otherwise never known: a country of small towns, front porches, rustic cabins, lumberjack breakfasts, church dinners, and roadside barbecues.

There was an outing on a lake in Olney, Illinois ("Home of the White Squirrel"), with my wife's uncle Herb. On a hot, midwestern summer day, we caught and cleaned a half dozen largemouth, then took them home to the family gathering for BBLT sandwiches: bacon, bass, lettuce, and tomato.

On assignment for *Field & Stream* in the limestone-rich meanders of the Ozarks, I spent an afternoon with a World War II army sharpshooter who had returned to Jackson Hollow, in Missouri, where he raised bluetick coonhounds. His dad had pulled similar sniper duty for Pershing in the Ardennes Forest in the First World War. They told me how they guided anglers on the White River before it was dammed. In those days, major league baseball teams barnstormed their way north, playing exhibition games after spring training. Some of the players took the opportunity to float the White for smallmouth. Father and son proudly reported that Stan "The Man" Musial was a regular client.

A few years later, I made a pilgrimage to Michigan's Upper Peninsula to visit with John Voelker, whose pen name was Robert Traver. I have seen his "Testament of a Fisherman" displayed on the walls of fishing cabins as reverently as a needlepoint embroidery of the Twenty-Third Psalm. While we were fishing for brook trout in the humid stillness of nightfall on Deadman Lake, no sooner had an ungainly *Hexagenia* emerged with great commotion than my fly was eaten by a jaws-wide-open smallmouth: my first encounter with *Micropterus dolomieu*.

On another assignment I drove down from Columbia, South Carolina, to the St. Johns River in North Florida and stopped to fish for bass and bluegills on a Georgia pond. It was owned by the only man I ever met with the first name of Mary. I caught a plump bass that fed under a limb where a cottonmouth lay coiled.

While I was researching Cajun fish camps in western Louisiana, an angling chef took me to Two O'clock Bayou. He enlightened me that Cajuns consider a bayou to be "something larger than a stream and smaller than a river." We caught pound-and-a-half largemouth that he blackened on a cast-iron skillet and served with a side of dirty rice, both dishes thermonuclearly spiced.

In short, bass have been my tour guide to an America that is as far a cry from my Brooklyn neighborhood as butterflies are from bacon.

For introducing me to bass on a popping bug, and much else in life, I am indebted to Jack "Bass" Allen. An eastern Oklahoma boy who grew up fishing for bass with Peck's Poppers, Jack later ran a bonefish lodge on Andros Island but, as he once told me, "I must have caught twenty thousand bonefish on a Pflueger Medalist, and I got tired of it."

He eventually settled in Fort Lauderdale and specialized in fishing for largemouth in the freshwater Everglades south of Okeechobee, using poppers as he had in his youth. "I think bass bugging is such a satisfying way of fishing," he would say. "Forget about all that pound-for-pound and inch-for-inch stuff. I get excited by the take, and there is nothing like the topwater take of a bass."

I fished with Jack for forty years. We used light flyrods suited to trout fishing, which he over-lined to cut down on false casts and to turn over air-resistant bugs. He fashioned his poppers from white Styrofoam that he salvaged from the necks of household cleaning containers. With his car-top canoe or trailering his no-frills johnboat, Jack drove us in his sun-faded Pontiac to the most remote spots in the Everglades. "I've never once encountered the need for a four-wheel-drive in the Glades," he said. With the windows rolled down and the cassette player blasting West Coast jazz, Jack regaled me with endless tales of the flyfishing life. He had a slow Oklahoma drawl and a born storyteller's cadence that I could have listened to forever.

And, oh, the bass we would catch! Not particularly big, but as he regularly observed, "Small but wiry." And then there was the abundant wildlife way back on the Miccosukee reservation where Jack got a free pass in exchange for the occasional stringer of bass. We saw herons, bitterns, kites, and coot. Flights of flamingos, blazing pink in the low light of the afternoon sun. On the shore, alligators galore, the occasional deer, and once, for just a second, the flash of

a Florida panther. We rarely saw another angler. On a calm day you could hear the soft kissing sound made by feeding bass, gathered in the cuts that opened onto sawgrass marshes.

Though Jack is gone now, whenever my cast falls apart, no matter what I am fishing for, I can hear his voice in my head: "Uh . . . Pete . . . slow down your back cast just a tad."

No one knew the freshwater Everglades as well as Jack "Bass" Allen.

Trout, bass, salmon, snook, or tarpon; there's nothing more thrilling than a fish on the line when it takes to the air.

PART V

ALWAYS LEARNING

If you flyfish long enough, you are bound to come across anglers who can teach you new things. This is a sport that can amuse, surprise, and delight no matter how many years you have been at it. Inevitably, you may discover a thing or two about yourself, too.

"Sun warmed the back of my neck, the air was dense with the sound of snapping grasshoppers and the smell of sage and pine, all mixed with the coolness rising up from the stream. Nothing then could have done me any harm."

GRETCHEN LEGLER

"Fishergirl"

Soft Corner on the Delaware by Galen Mercer

The Child Is Father to the Man

We were sitting by the riverbank at Wagon Tracks, a historic pool on the Beaverkill in the Catskill Mountains, cradle of American flyfishing.

"This wonderful sport," Lee Wulff said. Just three words, out of the blue, unprompted by conversation. In fact, not a word had passed between us for the last few minutes. But his phrase, as I understood it, summed up a feeling about flyfishing at the core of who he was, a sentiment that I share.

Lee was in his late seventies at the time, the most legendary American flyfisher of the twentieth century. We had fished awhile that day and caught a few trout, then laid our rods on the bank, picked up some flat stones, and skipped them across the surface of the pool, just like young boys. Four hops counted as a good throw. For a flyfisher, being on the water can make you feel like nine-year-old you—the child who lives on inside you no matter how many years have passed. You may take pleasure in different things, but the sense of delight remains as it was.

Ask me why I like a particular thing and I will respond with how it stacks up against flyfishing. I will drive long distances, sleep short hours, eat cold hamburgers, endure drenching rain, but once I am on the water, with a flyrod in my hand, I'm never unhappy.

When the day is fine and a few fish cooperate, the world seems completely right. There is no *why* to this; like faith, like love, it requires no proof. It just is—a species of sorcery that transports you out of the workaday world and into the place where dreams and visions arise.

For many, that scene is on a stream—a day in May, when pale mayfly wings flicker in the shafts of sunlight coming through trees in their first green blush. In my case, I tie on a dry fly, a Pink Hendrickson, size 14, devised for the Beaverkill a century ago by Roy Steenrod, a postal clerk in nearby Liberty, New York, and a tying disciple of the father of Catskill flyfishing, Theodore Gordon. When that hatch is on, the cold months of winter and April's fickle weather are banished from memory, and another year begins in earnest.

Lee Wulff Fishing the Serpentine River, 1940 (Photographer: Unknown).

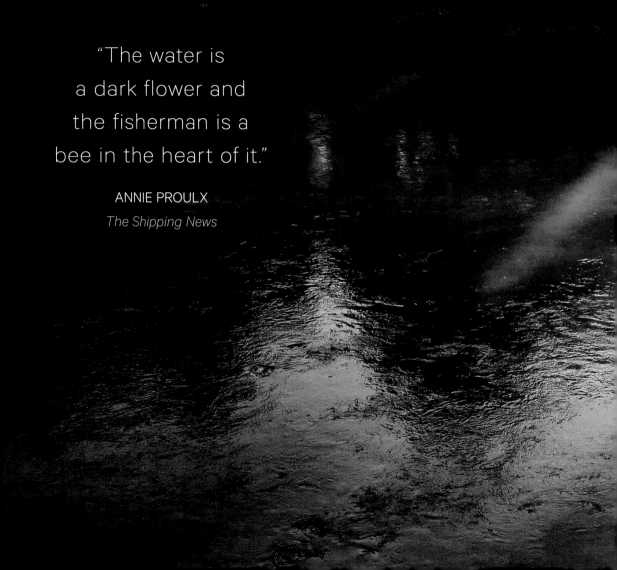

"These moments when the fly passes above the head of a trout are wonderful and excruciating. This is the same rush of excitement gamblers feel when they scratch off a lottery ticket. They don't really think they're going to win—they know better—it's the chance of winning they're addicted to."

DAVID COGGINS

The Optimist

The Pot No Longer Calls the Kettle Black

After so many years at this game, you would think that one might have outgrown the Christmas morning level of overexcitement that is occasioned by every fishing outing. It is a common affliction among anglers, and, like the equally common cold, it has no known cure. I once drove from Ketchum, Idaho, to Silver Creek, wadered up, and opened my rod case, only to discover that I had forgotten the rod. Miraculously, just as I was about to give up, a friend from New York City pulled into the parking lot and had an extra 5-weight on hand.

That Silver Creek slipup was not nearly as bad, though, as the time I started the drive home after an afternoon of striper fishing on Jamaica Bay. I was running a little late, so I quickly shucked off my overalls and hopped into the car. As I accelerated onto the Brooklyn Queens Expressway, my mind drifted off into a replay of that afternoon's ferocious top water eat of a big bass. Suddenly, I was yanked from my daydream by a rattling *clunk* on the roof of my car.

Glancing at the rearview mirror, I looked on in pain as my rod pinwheeled off my Subaru and under the oversized tires of a pickup truck. The rod was a favorite that I had used for years. It weighed next to nothing and cast like a laser. Losing it felt like a world-ending tragedy. This all happened rather quickly, but in memory I see it in agonizing slo-mo, visualizing it from every angle.

Served me right, too; for nearly fifty years I had regaled fishing companions with the anecdote of a friend—a superb fisher and an amazing writer—whom I had uncharitably described as boneheaded because of a similar mishap in the Catskills in 1976. We had just arrived back at our motel after fishing the Willowemoc, where my pal had given me my first fly-casting lesson. No sooner had we popped open a few beers than he realized he didn't have his rod. He recalled leaving it on top of his car after de-wadering. We retraced our route to the creek, but there was no rod by the roadside, nor at the pull-off.

Ever since my absentminded mishap on the BQE, I have omitted references to boneheadedness in my repertoire of fish tales. Being human, we are all at the mercy of such oversights. There's really nothing to be done about it: If you love to fish, it hijacks a part of your brain.

Similarly, during a fishing session, when a sense of beatific quietude sets in, even then there can be an occasional tremor in the Force. I might, for example, notice that my cast has lost some distance, or my double haul doesn't pull against the reel with the reserve force of a well-executed cast. "Damned rod has a dead spot," I tell myself. Just to check all the boxes, though, I'll inspect the rod, following the course of the fly line that I'd hurriedly threaded through the guides, and—sure enough—in my rush to get on the water, I'd missed a guide. I have learned to console myself with the assurance that such things wouldn't have happened if the thought of fishing didn't excite and delight me so.

"When you break it down,
there is no better way to catch a trout.
No other technique even comes close.
If there are rising fish and your fly selection
is on point, a fly is the best way to catch them.
Not a spinning rod, not bait, not nothing.
A fly will do the job better than anything."

STEVE DUDA
River Songs

Rachel Finn, Adirondacks guide and soulful artist, with her trademark stogie.

The Home Game

The sun had dipped behind the slope on the right bank of the West Branch of the Delaware. In the sheltering shadow of tall beech trees and spruce, wild brown trout fed on Pale Morning Duns (PMDs), their wings a pastel sunflower yellow. Predictably, for a hatch on this stretch of trophy water, a half dozen drift boats had anchored within casting range of the current seam favored by the trout.

"We're not fishing there," Al Caucci said.

No one knows the river better than Al. He and Bob Nastasi wrote the paradigm-shifting *Hatches* after years of scuba work and scientific study on the West Branch. They devised the Comparadun for the technical demands of fishing highly selective trout.

That day, I fully expected Al to find us a spot where we could shoehorn our boat in on the PMD conveyor belt. Instead, we pulled over to the opposite side of the river, where a foaming riffle flowed into pocket water. Swallows dipped and swooped to capture the hatching mayflies that emerged in clouds from the oxygen-rich current. Not the kind of water I dream about—I'm lovestruck by calmer, more visually seductive pools, dimpled with rise forms.

Al chose a Parachute Adams on 5X tippet—not a standard Caucci choice, but in the broken water the fly would be floatable and visible.

"There's a big trout here," he said, "but before you cast to it, I want you to put the fly somewhere else and tell me when you can see it."

It was hard to pick out the little white post on the Adams against the silvery jumble of the riffle. It took me a number of tries before I saw it well.

"Got it," I said.

"Okay, it's about a twenty-foot cast between you and the bank, slightly downstream."

I cast. A big trout flashed at my fly.

"Damn it," I said. "Missed my chance." I knew, or thought I knew, there are never second chances on the West Branch. Refusal equals point-set-match.

"He didn't feel the hook," Al said. "Go again. In this pocket water they're not so spooky."

I did as directed. The trout took. Startled the hell out of me. It was the biggest fish I had ever caught on that river, or any Catskill river: six pounds on a dry fly, a pattern that I would not have thought to use on a stretch I would have ignored, had my experienced pal not been calling the shots.

The lesson of that day: In fishing, as in many sports, it's better when it's a home game. You know the water, where every rock and riffle, every flat and shoal, has a name, and you have fished it in both fair weather and ugly days. If it's not your home court, most anglers will score more with someone who is intimate with that body of water. You will fish flies that never occurred to you, in places you would pass over. Fish with a friend, a family member, or a local guide until you get to know a bit more about the place.

A guide may seem like a luxury, especially when added to the money spent on airfare, lodging, and car rental for a long-anticipated fishing trip. But if you don't give yourself the best chance to catch fish, what's the point of spending all that money? Even if you can afford just one day of guiding on a weeklong trip, getting your cast pointed in the right direction by someone who has put in hundreds of days on a piece of water, in all conditions, taps you into knowledge that can be gained only through experience. It is the rare guide who hasn't taught me something, even on days when the fish or the weather—or both—don't cooperate. Knowledge born of experience can help even out the odds. Remember, for the fish it's even more of a home game.

Tip well; it will help your fishing karma.

Every angler who ever pulled over to check out a new pool knows this stop-and-look routine (and maybe took a sip or two).

Cruisin' all the Catskill streams

A broken-down Toyota to chase our dreams

Countin' the bugs on the windshield

And watchin' the swallows across the field

Tyin' flies on the old front porch

Let's follow the river around the bend

Guess we'll be home about half-past dark

This won't last forever, my friend.

CONNELLY AKSTENS
"The Yellow Mailbox"

West Kill in Spring by Steven Weinberg

Old

I'm not quite sure how our inner person—the me of it all—can transition from being a seven-year-old to an official old person. If flyfishing is your life's passion, it can, for the most part, forestall that reckoning, or at the very least keep you young in spirit. The ability to cast a fly often remains even as the knack for scrambling over jetties or wading thigh-deep in rushing water fades in your rearview mirror. Whenever I think about senior flyfishers, I am reminded of Harry Wilson's maiden float tube voyage on Idaho's Silver Creek.

If you are not familiar with Harry Wilson, he was the guy who started the Scott Fly Rod Company in the early 1970s. Legend has it that Harry would inspect J. Kennedy Fisher flyrod blanks right as they came off the assembly line and personally select the ones that met his demanding specs (most didn't). His Powerfly rods were the first flyrods I ever saw with spigot ferrules that allowed for the smooth transfer of force from one rod section to another. Loomis picked up on that design with the best casting—though maddeningly fragile—rods I have ever handled, the original GLX series. They were true cannons. My current workhorses, Orvis Helios models, are direct descendants.

Harry and I had struck up a phone friendship shortly after I began writing about the outdoors. One summer, I found myself for a few weeks in Ketchum, Idaho. Harry happened to be within driving distance, so I suggested that he come on over to float the Nature Conservancy water with me.

Harry was game, and a few days later I brought along an extra float tub that I had borrowed and met him at the put-in. He was, to my way of seeing the world in my midthirties, an old guy. When he stepped into the float tube and pulled it up around his waist, I am not being ungenerous when I say I was reminded of the dancing hippopotami in their tutus in Walt Disney's *Fantasia*.

A bit unsteady on his pins, Harry waded in. He made his way across the water to a line of tule reeds where trout were rising to a *Callibaetis* hatch. When he came within fifty feet of the rising trout, Harry uncorked a cast as straight and true as a Jedi light saber. No younger person could have thrown a tighter loop or with so deft a touch. He worked his line with the delicacy that one masters only after years on the water.

My friends, a posse of Sun Valley fish-aholics, held back from casting, happy to watch Harry pick off risers as he worked his way up the line. I remember thinking, *I hope I can do that when I'm his age*.

I believe I am now that age, and though I'm no Harry Wilson, my cast still gets there most of the time.

PART VI

THE ENDLESS NOW

What remains most strongly in memory is not the fish we have caught, the places we have been, the remarkable people we have met, nor the deep friendships we have forged. Instead, it is the feeling that washes over the angler like a wave and leaves behind a sense of soul-to-soul connection with the whole wide world.

"Sea birds scream of the carnage; their coarse signals carry for miles, attracting hundreds, sometimes thousands of their kin. Then the air above becomes part of the tumultuous mass—a sky filled with stripped feathers, the hysterical cry of anxious terns, the hoarser calls of the herring and black-back gulls, all diving, wheeling, hovering and heedless of any approach as they swallow the hapless bait fish whole whenever the prey are driven live from the sea or pick with their bills at the flesh fragments that rise in the wake of the stripers' feeding rush."

JOHN N. COLE

Striper

Birds

Ever since Noah, to his great relief, spotted a dove bearing an olive branch, birds have often been seen as harbingers of good fortune. For anglers, birds on the wing can be a tip-off that fishing prospects have brightened. The daredevil aerial display of swallows and swifts over a trout stream signals that something is hatching, which in turn means trout may be feeding. How the diving birds never collide as they dip and swoop in pursuit of insects in flight is a testament to the exquisite fine-tuning of evolution.

But it is not on trout streams that birds are the most important signifier for flyfishers. On salt water, birds are often the only sign of active, catchable fish. At first, the face of the sea may look as calm as the surface of a lily pond. Next, like the first quiet notes of a symphony that slowly builds to a crescendo, a few birds enter the scene, flying in high lazy circles . . . searching. From time to time a bird feints toward the surface, as if it has detected something. If bait starts to show, more and more birds flock in to feed until, at last, the water erupts with foam and fish. The birds, usually gulls, crowd together in deafening cacophony. They hit the water, squawking as they descend, as if to say, *Out of my way, bro!*

This is what you have been looking for: Game fish have corralled a bait school and driven it to the surface, a frothing mass of panicked alewives, rainbait, anchovies, and other small prey. They are easy pickings for the gulls, terns, and, late in the season, the gannets that soar a hundred feet over the water and drop like cannonballs on schools of luckless blueback herring.

A flyfishing outing during the great fish migrations of spring and autumn always starts with looking for birds. On a moving tide, game fish surround the bait school, drive it to the surface, or pin it against the shore, right in the suds, where waves die before being drawn back into the undertow. It is a scene of great carnage and mayhem. If this symphony began as a lilting pastorale, it is now a full-on Wagnerian *Götterdämmerung* or, as anglers have come to call it in an equally Teutonic term, a *blitz*.

It's hard to keep your wits about you or to know where to cast in the midst of 360 degrees of chaos, all around, above, and below. You cannot help but feel intoxicated at the sight of cresting waves revealing stripers, bluefish, or false albacore. They tumble in the curls of the thumping breakers that pound the shore with a rhythmic *whomp whomp*. If you don't catch something on every cast, it's because the fish have so much bait upon which to feed, or the ravenous game fish may be so tightly massed that your fly lands on their backs. The water turns dark red with blood mixed with chunks of half-eaten bait. The action is so furious that you have to remind yourself to exhale. Such encounters quicken the pulse, shorten the breath, and ignite the instinct that animates every predator, humankind included.

"Arriving there, we found the tide almost flood, with the water perfectly smooth and very clear and about a foot deep up at the mangrove roots. Here and there at a little distance, we could see splashes."

ZANE GREY
Bonefish

Snake Cay Creek by C.D. Clarke

Sometimes the panorama of a bonefish flat seems to go on forever.

The Light

When the sun peeks out from a puffy cloud, it lights up a bonefish flat like a smile on the face of a child. Other fishing scenes conjure their own emotions: serenity at moonrise over a lake's still waters; anticipation as morning mist lifts on a trout stream; the rush of adrenaline as stripers blitz in the curl of a breaking wave. A bonefish flat in full sun, however, overwhelms with a sense of immensity and infinitely varied colors. The sand catches the light so that the seascape is blindingly white in some places, warmly golden in others. Shadowy tracery marks channels where fish can hide or lurk. Green patches of eelgrass and turtle grass sway in the tide. The water is by turns aquamarine, shading into sky blue, and shot through with veins of amber before becoming no color at all, clear as air.

Contemplate this immense vista for a short time and you'll begin to pick out subtle signs of creatures going about their day on the flats. They are well camouflaged, so it is only by noticing their movements that you can begin to recognize them. A manta ray and the slo-mo flap of its wings sends up clouds of sand as it feeds along the bottom. The glint of a scale catching the sunlight reveals that what appears to be driftwood is, in fact, two barracudas lying in wait. Here and there, a patrolling sand shark turns with the flourish of a race car in a power skid.

At some point, you will see the torpedo shapes of bonefish and you will lock in on the wisp of black at the tips of their fins. Even if you have fished for bones many times, it always takes some reacquainting before you can pick them out in the vista. But once your vision has adjusted, you can reliably spot them on a hundred-acre expanse. Any discourse with the world, then, will be like a prayer, beckoning the bonefish into casting range.

The fish themselves are in no hurry. There may be six of them. Or fifty. Or only a solitary cruiser. You will get just one cast, and it must be placed in front of the bonefish's line of march. You wait, calibrating second-by-second course corrections before you launch so that your fly lands where it has a chance of enticing the fish. Once it does, one of three things can happen. First, an errant cast might startle fish and send them sprinting to deeper water. Second, an imprecise cast can elicit no reaction at all from the bonefish, tempting you to pick up and try another cast. Don't bother; I have never caught a bonefish on the second cast. At best it will turn the school as it moseys off with taunting insouciance.

The third and most hoped-for possibility is that one fish will deviate ever so slightly from its course and grab the fly. You'll strip strike and raise the rod as the bonefish rockets into a run across the flat. At that instant, your world, which had irised down to a narrow tunnel between you and the fish, opens up wide to take in the whole sun-splashed expanse and you may have the feeling that you, too, are filled with light.

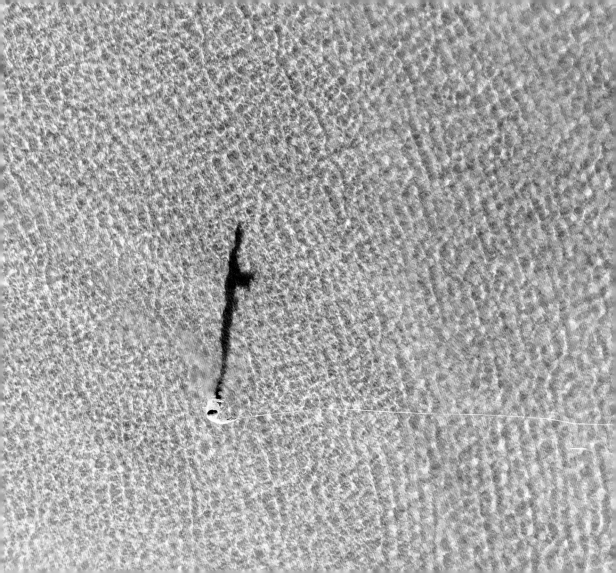

The changing moods of light on the flats.

High in the Andes

Although drinking and flyfishing are both pleasures, drinking while flyfishing is another story. Apart from the hazards of being tipsy when wading a river, hopping over jetties, or driving a boat, your cast is liable to lose some finesse. A drink *after* fishing, however, often feels well earned. For years I would finish a fishing session and trudge a thirsty mile back to the car from Long Island's Connetquot River (the very same water where Daniel Webster caught his world-record brookie in 1828). On the way home, I'd always pull into the last gas station before getting on the highway and buy a long-neck Bud and a bag of spicy barbecue-flavored tortilla chips for the drive to town. Drinking and driving? "Well, it was just a beer, officer."

My rules for cannabis and flyfishing are less stringent. Basically, I never start out fishing high or even slightly buzzed. You're pretty much guaranteed to skip a guide or two while stringing up your rod or forget a fly box on the riverbank. Bottom line: If you want to get high while fishing, wait until you are in a groove. If things are going well and your cast is flowing, it can make a good day better; or perhaps it's more accurate to say it can make that day good in a different way.

Case in point: a lodge in the Andes where, along with a group of anglers, we fished a glacial lake and the river that flowed out of it. On our next-to-last night, two anglers reported that they had been

Top of the Canyon-Malleo River by Bob White

having a hard time managing the winds—a constant in Patagonia. On that day, however, the lodge's program of rotating beats had put them on a sheltered cove of the lake where large trout basked in the white sandy shallows. If you placed a big Chernobyl Ant within thirty feet of the fish, they would often cruise over and suck down the fly.

"Best fishing we've had on the whole trip," they said. According to the plan, I was to be on that beat the next day. But, since I was a guest of the lodge, and these two guys had traveled five thousand miles and spent a lot of money to have so-so fishing, I said, "Why don't you take my beat tomorrow?"

The fishing that day was very good at the lake. The two anglers were ecstatic. The guests and guides had a farewell asado that night, and one of the guides, whom I'll call Jeremy, told me his sister, a bartender in Montana, was about to arrive. Now that the season was done at the lodge, they were going to fish for a few days. Would I like to join them? he asked.

All season, as gales blowing off the glacier whipped the lake, the only fishable spot had been that single white patch. Guides and anglers felt fortunate that Providence had provided this little honey hole. Wondrous to behold when we arrived at the lake, it was completely glassed over without a breath of wind! There were white sandy patches everywhere and the submarine-shaped shadows of big trout on each patch—browns and rainbows, many of them over twenty inches. Was it a karmic reward for my sacrificing that spot the day before? It sure felt like it.

We caught trout after trout. I was in rhythm, Jeremy and his sister were in rhythm, the trout were in rhythm. When she produced a big fat joint, I saw no reason not to smoke. Damn, it felt good! We would cast, then pause as the fly settled, wait until the fish noticed (they always did), wait some more as the trout lazed over to the fly, and then, as they softly took, raised our rods just as gently and the fish came tight . . . every time.

We were blissed out, copacetic, blessedly entranced—in other words, happily high. We caught fish until we felt we'd had our fill. I clearly remember hearing John Fogerty singing "Down on the Corner" while we fished, so they must have brought a boom box . . . or maybe I imagined it.

The vibe when it all goes well.

"A canoe trip has become simply a rite of oneness with certain terrain, a diversion of the field, an act performed not because it is necessary but because there is value in the act itself."

JOHN McPHEE
The Survival of the Bark Canoe

"It is impossible to grow weary of a sport that is never the same on any two days of the year."

THEODORE GORDON

The Persistence of Memory

There is no fishing conveyance quite so graceful as a canoe. While anything that puts you on fish is great, the way that a canoe glides through the water, carving out the quietest V-shaped wake with every stroke, is a sublime unity of form and function. The rings made by droplets falling from your paddle are small wrinkles in a serene still life.

To be sure, when you are on the sea and a blitz explodes, you can work yourself into a lather, getting a skiff up on plane on the way to a cloud of screeching gulls. Or, even without a craft, wading into rushing water to get in position for a well-considered cast can trigger an equal jolt of adrenaline. But for easing into tranquility, canoeing is hard to beat. Some of the fishing experiences most deeply engraved in my spirit have their origin not in furious action, but in time spent in a canoe.

There was, for example, a trip to the Grasse River, just north of the Adirondacks. It's a nice smallmouth stream that flows into the St. Lawrence. We were a party of four: my longtime fishing buddy, Connelly Akstens (a licensed guide and a Shakespeare professor); Peter Hornbeck, who designed our canoes (each weighing just eleven pounds); and my daughter Lucy, who was fourteen years old at the time. The journey from the cabin my family had rented near Lake Placid was a drive of a hundred miles or so through a halcyon America; most of the houses had steeply sloped tin roofs designed

Still life with drift boat.

to shed heavy snowfalls before the weight of their drifts caused any damage. Here and there we would pass one of the nineteenth-century Great Camps with their signature twiggy design.

It was a gorgeous summer day, easy paddling, few rapids. The company, the scenery, the beams of sunlight coming through trees in full leaf, and the whisper of canoes moving forward with each stroke felt like an ode to the fullness of summer. Though we didn't have lights-out fishing, just enough bass ate our Bead Head Wooly Buggers that we could count the trip as a success.

What stands out, though, is the image of Lucy in her canoe—the rod in front of her pointing downriver, the paddle resting crosswise over her lap, the sun bright, the air clear, and the look on her face, placidly soaking up the sun with a serene half smile. It stirs a memory I have of coming upon George Caleb Bingham's *Fur Traders Descending the Missouri*. It amazed me, as a young boy, whenever I visited the Metropolitan Museum of Art with my grandpa Jan. I always made a point of seeking it out (but only after visiting the mummies first). On that day on the Grasse, I felt as if I were in that painting, casually puffing on a corncob pipe, oiling my flintlock. Lucy, like those ancient rivermen, seemed very much aware of being on the water yet thoroughly at ease, as if to say, *No need to do anything at all. The river will bear me along in this endless now.*

Is it the size of the fish or the size of the world that captivates us so?

"The season is ended. There was not enough of it; there never is."

NICK LYONS

Photography and Art Credits

American Museum of Fly Fishing and
    Joan Wulff, page 117
Andrew Burr, page 37
Andy Anderson, pages i, 44, 52,
    62–63, 72, 153, 159
Bob White, pages 98, 150
Brendan McCarthy, page 91
Brian Grossenbacher, pages 54, 71,
    108–109, 120
Capt. John Mauser and Jones Brothers
    Marine, page 139
C.D. Clarke, page 142
Clay Banks, pages 26–27 (center)
Daniel Coimbra, page 88
Darcy Bacha, pages 8, 25, 32, 68–69,
    110–111
David Lambroughton, pages 2, 39
David Reilly, pages 17, 157
Dylan Schmitz, pages 102, 144–145
Galen Mercer, page 114
Jared Zissu, pages 18–19, 51, 82–83, 126
Jeremiah Watt, pages vi–vii

Jess McGlothin, pages x–1, 112
Joel Stoehr, page 101
Keith Meyers for the *New York Times*,
    page 96
Lael P. Johnson, pages 56–57
Nick Price, pages 20–21, 30, 43, 148,
    149, 162
Nigel Nunn, pages ix, 5, 35, 61, 95
Noah Rosenthal, page 85
Oliver Rogers, pages 136–137
Pat Ford, pages 64–65, 80–81
Paul Svensson, page 26 (left)
Peter Kaminsky, page 75
Phil Street, page 27 (right)
Rex Messing, pages 48, 58, 130, 161
Stan Wood, page 107
Steven Weinberg, page 132
Tim Romano, pages 7, 11, 46–47, 77,
    129, 154
Todd Towle, pages 28, 42, 125
Tosh Brown, pages 78–79
William Hereford, pages 12, 118–119

Acknowledgments

I am grateful to Danny Cooper for dreaming up the idea for this book and for shepherding it every step of the way. Suet Chong for her elegant vision. Lia Ronnen, whose support has always been from the heart. Laurel Robinson, whose copyedits make all the difference. David Black, a true mensch. My wife, Melinda, who read every word over and over until the book was about as good as it was ever going to be. Nick Lyons and Stephen Sautner for their close reads. Henry Hughes for his blue pencil and his encouragement. Finally, to the photographers and fine artists, whose work appears in these pages: They inspire me.

PETER KAMINSKY is one of America's leading angling journalists and authors. His Outdoors column appeared in the *New York Times* for thirty-five years. He has been a contributing editor at *Field & Stream*, *Sports Afield*, and *Outdoor Life*, and was managing editor of *National Lampoon*. His angling writing has also appeared in *Fly Fisherman*, *Condé Nast Traveler*, *Smithsonian Magazine*, *GQ*, *The Field*, and *Anglers Journal*. Among his fishing books are *The Catch of a Lifetime*, *The Moon Pulled Up an Acre of Bass*, *American Waters*, and *The Flyfisherman's Guide to the Meaning of Life*. An avid home cook, he wrote *Bacon Nation*, *Pig Perfect*, and *Culinary Intelligence*. He has coauthored nineteen cookbooks including *Seven Fires* with Francis Mallmann, *The Elements of Taste* with Gray Kunz, and *Ultimate Tailgating* with John Madden. He was The Underground Gourmet for *New York* magazine for four years, and his food writing has often appeared in *Food & Wine* and the *New York Times*. As writer and executive producer, Kaminsky is a creator of The Kennedy Center Mark Twain Prize for American Humor and The Library of Congress Gershwin Prize for Popular Song. He lives in Brooklyn, New York.